TEXT BOOK

ON

AERIAL GUNNERY.

LONDON:

The Naval & Military Press Ltd
© 2008

In reprinting in facsimile from the original, any imperfections are inevitably reproduced and the quality may fall short of modern type and cartographic standards.
Printed and bound by CPI Antony Rowe, Eastbourne

CONTENTS.

Part I.
	PAGE.
Introduction	3

Part II.
| Instructions for using Ring Back-Sight and Wind-Vane Foresight (Norman Pattern) ... ... | 9 |

Part III.
| Ring Sight for Fixed Machine-Gun on an Aeroplane ... | 11 |

Part IV.
| The Aldis Optical Sight ... ... | 12 |

Part V.
| Instructions for Using Model Aiming Aeroplane ... | 14 |

Part VI.
| Instructions for Using Hythe Gun Camera: Mk. II. ... | 17 |

Part VII.
| Shooting Practice at a Picture Target ... ... | 22 |

Part VIII.
| Notes on the Use of Various Practices ... ... | 28 |

MACHINE GUN SHOOTING IN THE AIR, USING RING SIGHTS.

Part I.—INTRODUCTION.

1. In all the following it is assumed that a machine gun is used with Mark VII ammunition or tracer.

Difficulty of taking a Steady Aim in the Air.

2. Experience has shown that a gunner firing from an aeroplane cannot hold his gun very steadily upon a mark which is passing him, but that, even after considerable practice, he will as a rule scatter his bullets over a fairly wide area around that mark. This is principally due to two distinct causes—the difficulty of traversing a gun steadily so as to follow a moving object, and to small angular movements of the aeroplane, due to unsteadiness of the air.

3. The scattering, due to want of rigidity in the mounting, is small compared with that due to the above two causes; for it is easy, when firing from an aeroplane on the ground at a fixed mark, to obtain groups much more concentrated than can be obtained in the air under average conditions.

Size of Group under various Conditions.

4. If the air is at all " bumpy," as it nearly always is within 1,000 feet of the ground, the scattering due to bumps is greater than that due to any other causes. The group of shots round an object aimed at under these conditions has seldom been less than 50 feet across at 250 yards range; and this size of group has often been greatly exceeded, even by an experienced gunner.

5. In perfectly steady air, however, such as rarely exists near the ground, but is often found at high altitudes, it is quite possible to concentrate nearly all the shots fired into a 20-foot circle at 250 yards range, provided that the gun has not to be traversed (when, for instance, two aeroplanes are flying side by side at the same speed). If, on the other hand, the gun has to be traversed at all rapidly, the shots, even under the steadiest conditions, will not be contained in less than about a 30-foot circle, except when fired by a very good gunner.

6. In general it will be assumed in what follows that, under the conditions usually met at high altitudes, a practised gunner will spread the majority of his shots over a circle of about 30 feet diameter, or 15 feet radius around his point of aim, when the range is 250 yards. The size of this circle will vary roughly in proportion with the range, except when it is necessary to traverse the gun very rapidly, as is often the case at short ranges.

Allowance for Speed of Gunner's Aeroplane.

7. When a gun carried in an aeroplane is pointing across the line of flight, the motion of the aeroplane carries the bullet ahead of the spot at which the barrel is pointing. Thus, to hit an object fixed in the air, when flying past it, the gun must be pointed behind the object, and therefore the sights must be tilted so as to point ahead of the gun.

8. In a machine flying through the air at 100 m.p.h., the allowance for this effect—called the gunner's speed allowance—amounts to 46 feet at 250 yards' range, and at other ranges is in proportion to the range.

Gunner's Speed Allowance Automatic.

9. In the sights now issued the gunner's speed allowance is made automatically, either by carrying the foresight on a wind-vane so that it is offset by a certain amount when the gun is pointing across the aeroplane, or, in the case of the telescopic sight, by mechanism which tilts the telescope in relation to the gun.

10. In most existing designs the sight is adjusted for one particular speed, and the allowance is correct for that speed only. This is chosen so as to be near the average speed of the aeroplane on which the sight is used.

11. When the aeroplane is not flying at this speed the allowance will be incorrect, but the error will usually be such that it is covered by the scattering of the individual shots. For instance, if the sight is correct for 80 m.p.h. and the aeroplane dives at 100 m.p.h., the error in aiming at 250 yards will be about 10 feet. Hence with a group 30 feet in diameter, the target will still be within the group if the gun be otherwise correctly aimed.

12. For the present, therefore, the gunner will be taught to neglect the changes in his own speed, for he has quite as much as he can manage to allow for the enemy's speed in the time usually at his disposal. It is possible that future sights may be devised to allow automatically for changes in the gunner's speed, but these, if they come, will only increase the accuracy of shooting without in any way altering the training required.

13. The gunner, therefore, when using these automatic sights, takes no account of his own speed, but aims exactly as he would were he firing from the ground.

Sights for Fixed Gun.

14. In fixed guns firing straight ahead, no allowance for own speed is necessary, since the speed of the aeroplane is merely added to that of the bullet and gives it no sideway drift.

15. On fixed guns firing at a small angle with the aeroplane a fixed allowance for gunner's speed only is required. With any fixed gun, therefore, only fixed sights are provided, and all wind-vanes and other moving mechanism are dispensed with.

Allowance for Speed of Enemy Aeroplane.

16. The automatic allowance for the gunner's own speed enables him to hit a target stationary in the air. When aiming at a target which is itself moving through the air it is necessary to make an additional allowance by pointing the gun ahead of the target, because the target will move through the air during the time that the bullet takes to reach it. This is called the enemy's speed allowance, and is rather greater than the gunner's speed allowance for the same range, owing to the fact that the bullet slows down during its flight.

17. If the machines are flying in opposite directions the allowance for gunner's and enemy's speed add up, so that if each aeroplane is flying at, say, 100 m.p.h., the total allowance that would be necessary, if no automatic sights were used, would be nearly 100 feet at 250 yards' range.

Method of Allowing for Enemy's Speed.

18. With either the ring and wind-vane sight, or the Aldis optical sight, the correct sighting line to allow for the gunner's own speed is arranged to pass through the centre of a ring of such a radius that an aeroplane, flying directly across the line of sight at 100 m.p.h. air speed, would pass from the edge of the ring to the centre in the time taken for the bullet to reach the aeroplane. This is true with sufficient accuracy for all ranges at which air shooting can be profitably carried out.

19 If, therefore, the gunner should see a target aeroplane flying at 100 m.p.h. directly across his line of sight—*i.e.*, if the body does not appear foreshortened—he must lay the gun so that the vital parts are on the edge of the ring and the *body pointing towards the centre.*

20. Should he see the aeroplane flying partly towards or away from him—*i.e.*, if the body does appear foreshortened—he must place the vital spot somewhere within the ring, at a distance from the centre that depends on the amount by which the body is foreshortened.

21. Should the body be pointing straight towards or away from him, he will place it in the centre of his ring and aim point blank.

22. In any case it is absolutely essential that the body should point towards the centre of the ring.

Allowance independent of Range.

23. The gunner should be taught that the correct position of the aeroplane in the ring for a given speed depends upon the foreshortening of the body only and *not upon the range*, for although theoretically there is some difference in allowance at different ranges, this difference is insignificant.

Early Training to Assume a Fixed Enemy's Speed.

24. For the sake of simplicity, all the early training should be done on the assumption that the enemy aeroplane is moving at the correct speed for which the ring is intended, *i.e.*, 100 m.p.h. with the sight issued. The training should not be complicated with additional allowances for probable variation of speed due to diving, climbing, etc., until the gunner has become thoroughly accustomed to making his allowance correctly on the simpler assumption of a fixed speed.

Necessity for continual Training in making the Allowance.

25. The estimation of the amount the body is foreshortened, and the operation of laying the gun so that the target aeroplane appears at the corresponding distance from the edge of the ring and with its body pointing to the centre must, therefore, form the fundamental part of the training for all gunners intending to use these sights in the air. In most aerial fighting the target is altering its appearance with bewildering rapidity during the entire fight, so that it is essential that the gunner should have developed by constant practice, the habit of laying his gun correctly for the momentary appearance of the aeroplane and of following the target's movements whilst he is firing, so that it continually occupies the correct position in the ring. Not until this can be done almost as easily and well as the gunner can lay a sight directly on a target, will he be of much use in an actual fight. If he has to think much before laying the gun he will fail to get in his shots in time.

26. The gunner, therefore, must practise continually, both during training and in the intervals between fights, at estimating this allowance as rapidly as possible, and as it is usually impossible to get sufficient flying to give him the amount of practice required, he must work continually on the ground, either at the aiming model or at the picture target, or by any other means which may from time to time be devised to give him a realistic view of a target aeroplane and an indication as to whether he has made his allowance accurately.

Necessity for generality in the Training.

27. During the whole of the training great care should be taken never to let the gunner practise for long at any one range, or at any one aspect of the aeroplane. If he is allowed to do either of these things he invariably forms habits of making the allowance by some means which is peculiar to the range or aspect at which he is working; since both range and aspect are always rapidly altering in a real fight, this is the very worst thing he can do.

28. For instance, if, when using the aiming model, the range is not altered, the gunner rapidly finds out that he can score a hit by aiming so many body lengths ahead, and instinctively forms the habit of doing this, to the complete ruin of his shooting in the air.

29. Again, if the picture target on the range is not altered the gunner will soon find that by laying his ring on some part of the machine—say, the tail—he can score a hit.

30. In actual fighting the target may be seen from any point of view whatever, and on account of the gunner's own speed may appear to be moving in any direction, regardless of the direction in which its body is pointing. For instance, if the enemy aeroplane is approaching from one side, it will appear to be moving rapidly sideways in the direction of the tail of the gunner's machine, even although its body may be pointing directly at the gunner. Or, again, should it be flying alongside and in the same direction as the gunner, it will appear to be travelling backwards if it is a slower machine. The gunner, on the other hand, has to learn to make his part of the allowance depend only upon the direction in which the body is pointing, so that any practices he may use should teach him to do this from whatever point of view he sees the target and wherever it may appear to be moving. For these reasons, in practices at moving models, the models should be constantly varied and be made to move so that the aeroplane does not appear to be moving along its own body.

Also, in practices at picture targets (*see* Part VII), not only should the pictures be constantly changed, but care should be taken that the direction of the body of the aeroplane on the picture is varied. Moreover, the gunner's cockpit should be rotated sometimes one way, sometimes the other, or the rotation may even be stopped and reversed while the firing is proceeding.

Range at which to Open Fire.

31. The maximum range at which fire may be profitably opened must depend very largely on individual skill, and on such circumstances as the relative speed and aspect of the two aeroplanes. It is impossible at present to lay down anything very definite on this point, nor will it be possible until experience has been gained of the actual results obtained with the new pattern

sights by men thoroughly trained in their use. Sufficient information has been obtained, however, to show that under some circumstances fire may be opened with advantage at 300 yards, but that that range should rarely, if ever, be exceeded. On the other hand, at ranges of 100 yards or less the relative position of the aeroplane may be changing so rapidly that the average man cannot use his sights with any effect, but must rely on manœuvre and tracer bullets. Thus it would appear probable that very fast and handy scout machines will try to close with the large two-seaters, so as to get full advantage from their power of manœuvre, while the large machine must rely on accurate gunnery at long range to keep off the attacking scout. If the latter can get within 100 yards he will usually have the big machine at his mercy. On the other hand, the big machine having more guns and field of fire has the advantage at longer ranges between 100 and 300 yards.

32. For the present, therefore, observer pupils should be taught to reserve all fire to within 300 yards and to try to get in as much as possible at about 200 yards. The range can be judged by the apparent size of the target within the ring. Considerable practice is required, however, before this can be done accurately, because the size of an aeroplane at a fixed range will appear to vary considerably when it is viewed from different aspects. The best training for this purpose is to carry out aiming practice at the picture target and at the aiming model, at all ranges up to the equivalent of 300 yards, but never beyond, so that the target will look unfamiliarly small at ranges above this distance. Particular attention should be paid to this point when using the gun camera.

33. The above considerations do not, of course, apply when attempting to affect a surprise attack, or when from any other cause it is possible to get in shooting at close range with the two machines moving practically in the same direction, so that the relative velocity is small. In this case the shorter the range the better from the gunner's point of view.

Use of Tracers.

34. Tracer bullets give but little indication of the allowance necessary for relative velocity, and should therefore be disregarded when the sight can conveniently be used. At close range, however, under 100 yards, they form a good substitute for a point blank sight, and can often be used with effect when it is impracticable to use the sight, owing to want of time, or to the difficulty of getting the head into position behind the sight.

To face page 9.

Fig. 1.

PHOTOGRAPHS ILLUSTRATING RING BACK-SIGHT AND WIND-VANE FORESIGHT (NORMAN PATTERN).

Part II.—INSTRUCTIONS FOR USING RING BACKSIGHT AND WIND-VANE FORESIGHT (NORMAN PATTERN).

Description of Sight.

1. The illustrating photographs (Fig. 1) show the sight in place on the gun.
2. The foresight A consists of a bead H, which is moved by the wind-vane K so as to allow for the speed of the gunner's aeroplane, no matter in what direction the gun is pointed.
3. B is the ring backsight, by means of which the necessary allowance is made for the enemy's speed.
4. The sights are carried on the gun by means of the adapters C and D, from which they can be easily removed and replaced without disturbing the accuracy of setting.
5. The adapter C is jammed on to the taper part of the barrel by means of the nut J, and, as there is no key to prevent the adapter from turning upon the barrel, this nut must be very firmly screwed home. Special nuts for this purpose will be provided; until these can be issued the barrel mouthpiece from a land gun may be used.
6. The adapter D grips the front band of the small radiator casing as shown in the photograph, the yoke being clamped on the rear band.
7. A fixed bead foresight may be put in place of A to check the setting of the adapters, and for use either with the Hythe camera gun or the aiming model.

Adjustment of Sight.

8. Lateral adjustment only is needed with these sights, as the positions of the holes for the pins E and F are such that the sights, when aligned laterally and upright on the gun, are also correct for elevation.
9. Lateral adjustment can be affected by loosening the adapters and tapping them round until both sides are upright on the gun, with their stems parallel. It should be possible to do this with sufficient accuracy by eye, but if a check is required, replace the foresight A by the fixed bead and adjust by looking through the barrel of the gun at a distant object in the ordinary way. The centre of the ring backsight and the fixed bead should then line up on the same distant object as the barrel of the gun.

Method of Using the Sights.

10. Two wind-vane sights are issued marked respectively " 80 m.p.h. at 18 inches " and " 100 m.p.h. at 18 inches."

11. When the sights are correctly adjusted the centre of the ring backsight B and the bead H give the correct sighting line for the bullet to hit an object stationary in the air, when fired from an aeroplane moving at the marked speed (80 or 100 m.p.h.) through the air. This is true, no matter in what direction the gun is pointing out of the moving aeroplane, because the wind-vane sight A automatically applies the necessary correction for the aeroplane's speed. The gunner therefore has no allowance to make for his own speed and need not consider it at all when firing.

12. When firing at an aeroplane which is itself moving through the air, however, the gunner must make a further allowance so as to aim at the spot in which the target will be when the bullet arrives there. For this purpose the ring backsight is provided; this is marked " 100 m.p.h. at 19 inches," and is correct for use against a target moving at 100 m.p.h. through the air when the eye is placed at 19 inches behind it.

13. When aiming, first align the eye so that the bead H appears to be in the centre of the ring of the backsight B. The head must be held in this position throughout the firing, and must be kept approximately over the handle of the spade grip so that eye is about 19 inches behind the backsight. An error of 1 inch either way in this distance does not matter, and with practice it is not difficult to acquire the habit of placing the eye within these limits. This is best done by resting the chin or cheek upon the hand holding the spade grip, or upon the pad above the handle, if one is provided (see Fig. 1).

14. If a butt stock is used it should be cut down until the eye comes at the correct distance behind the sight, otherwise the eye will be too far behind the ring and the allowance will be suitable for some speed lower than 100 m.p.h.

15. With the head held in position as above, the target aeroplane must be placed within the ring of the backsight, in a position which depends upon the aeroplane's appearance, and which is described in detail below.

Correct Position of Enemy Aeroplane in Ring (see Fig. 2).

16. The position of the enemy aeroplane within the ring at the moment of firing must be such that it will fly into the centre of the ring by the time that the bullet reaches it. For this reason the first essential is that the body of the aeroplane must be pointing directly towards the centre of the ring.

17. The distance that the vital part of the aeroplane must be placed from the centre of the ring depends upon the angle at which it is approaching or moving away.

18. If it is flying directly towards, or away from, the gun, it must be placed exactly in the centre of the ring (A in Fig. 2).

To face page 10.

Fig. 2.

Photographs illustrating use of Ring Back-sight with Wind-vane Foresight (Norman Pattern).

B.E. 2c on 100 m.p.h. Ring Sight.

A B C D E F *Range.*

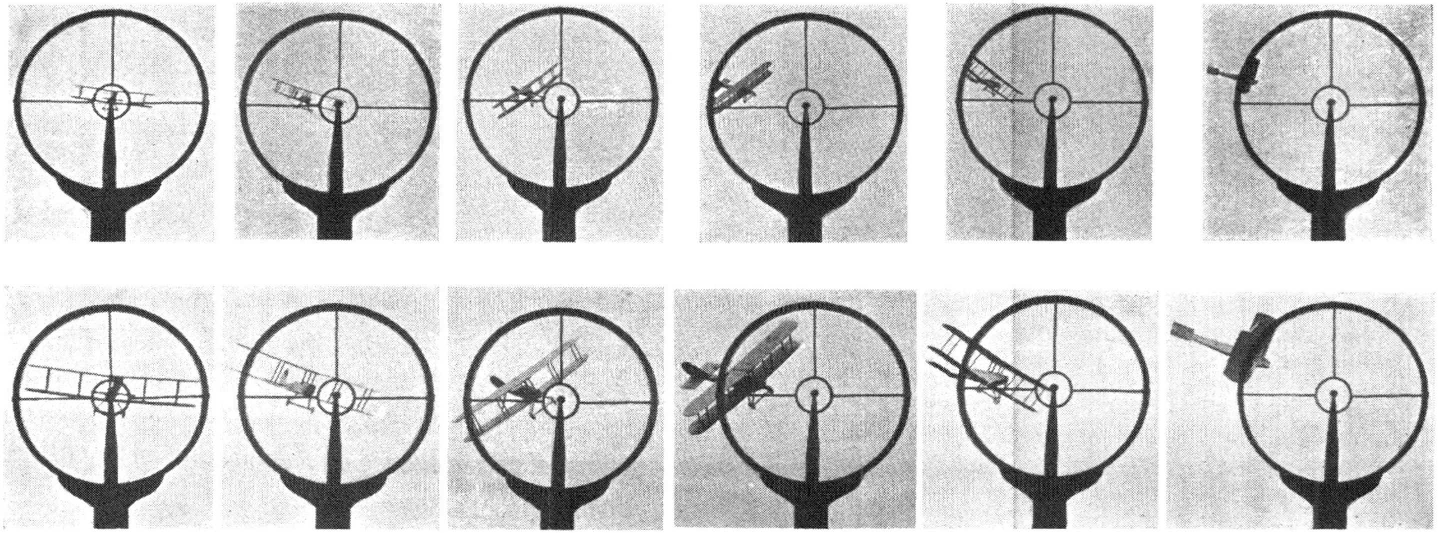

200 yds.

100 yds.

19. If it is flying directly across the line of sight it must be placed on the outer ring (F in Fig. 2).

20. If it is flying obliquely towards or away from the gun it must be placed between the centre and the outer ring, at a distance from the centre which depends on the angle of approach. This distance can, with practice, be judged with sufficient accuracy from the appearance of the aeroplane (B, C, D and E in Fig. 2).

21. The distance of the vital part of the aeroplane from the centre of the ring should depend only on the angle at which the target aeroplane is approaching or moving away from the gun and not upon the range, or apparent size of the target in the ring.

22. Care should be taken to avoid estimating the necessary allowances in body lengths of the aeroplane, since any system based on this method is correct at one range only and will be wrong at other ranges.

23. The correct use of a ring sight requires considerable practice to enable the gunner instantly to associate each aspect from which he may see the target with its correct position in the ring and to accustom him to making his allowance independent of the target's apparent size. Useful forms of practice are described in parts 5, 6 and 7 of these notes.

24. With some people the operation of laying the gun to allow for the enemy's speed, is most easily performed by making the enemy aeroplane touch an imaginary sphere, which just fits the ring in such a way that he is flying towards the centre of the ring.

This method of making the allowance is quite correct.

Part III.—RING SIGHT FOR FIXED MACHINE-GUN ON AN AEROPLANE.

Description of Sight.

1. This sight consists of two concentric rings 5 in. and 1 in. in diameter respectively, and four radial lines joining the inner and outer rings.

2. It can be used either as a foresight with a fixed bead backsight or a peep-hole, or as a backsight with a fixed bead foresight. In either case the radius of the outer ring is correct for use against an enemy aeroplane travelling at 100 m.p.h. when the *ring* is 38 inches from the eye. If placed further from the eye the speed allowance is reduced in proportion, thus at 42 inches it corresponds to 90 m.p.h. and at 34 inches to 110 m.p.h.

Use of the Sight.

3. The sight is used exactly the same way as the ring and wind-vane sight described in Part II. The head must be placed so that the bead appears in the middle of the inner ring, and the aeroplane manœuvred until the enemy aeroplane appears in the correct position in the ring to allow for its own speed. This position is found exactly as described previously.

4. The only differences between this sight and the ring and wind-vane sight are that the movable bead on the wind-vane is replaced by a fixed bead (this is because no allowance is necessary for the gunner's own speed) and that the ring is twice as large and twice as far from the eye (this is to allow the gunner twice the margin for error in the position of his eye).

Use of Sight with Fixed gun not Parallel to axis of Flight.

5. If the fixed gun is pointing at an angle to the direction of flight, the sight can be set up as follows:—

6. Fix the ring at the correct distance from the eye.

7. Erect a temporary bead so that the line joining it to the centre of the ring is parallel to the barrel of the gun.

8. Set up the permanent bead so that the line joining it to the temporary bead is parallel to the direction of flight and so that the distance between the two beads is equal to $\frac{1}{16}$th the distance between the temporary bead and the centre of the ring.

9. If the bead is a backsight the permanent bead must be behind the temporary bead, but if a foresight it must be in front.

10. Remove the temporary bead; the permanent bead is now correct for a speed of 100 m.p.h in the gunner's aeroplane.

Part IV.—THE ALDIS OPTICAL SIGHT.

Description of Sight.

1. The Aldis sight is virtually a telescope which does not magnify or diminish, and which, unlike an ordinary telescope, can be used with the eye several inches from the end of the tube.

2. When looking through this tube at a distant object, the effect is exactly as though one were looking through a napkin ring: the object appears the same whether it is seen through the tube or outside it.

3. Within the tube is a glass screen carrying the sighting circle, which is so placed between the lenses that on looking through the tube at a distant object the ring is seen with its centre on the

spot at which the tube is pointing, no matter where the eye is placed. If the eye is moved sideways the ring appears to move with it through the telescope, so that the direction in which the tube points is always towards the centre of the ring

4. The tube, when fixed rigidly to a gun, therefore, constitutes a sight which offers practically no obstruction to the view, and which shows instantly the spot at which the gun is pointing, without the necessity of aligning the eye upon a front and backsight.

5. The sight can with advantage be used with both eyes open, one eye sees the object and the circle through the tube, and one eye sees the object direct. The effect after a little practice is that the object is seen as clearly as though there were no sight at all, but that a circle appears in the sky, the centre of which is the point where the tube is aiming.

6. These tubes are supplied in two sizes—1-inch diameter and 2-inch diameter. The former is intended for use with a moving gun and the latter with a fixed gun. The special advantage of the large size is that the eye can be held further from the end of the tube, and has more freedom of movement, whilst still keeping the target within the field of the telescope.

Use of the Sight.

7. For use with fixed guns firing directly forward from the aeroplane, the sight must be fixed parallel to the axis of the gun. The centre of the circle will then always fall on the point that the gun will hit if that point is not moving through the air; in this case no allowance for the speed of the gunner's aeroplane is necessary.

8. If, however, the sight is used on a movable gun, means must be provided for tilting it to a small angle in relation to the gun, so as to allow for the speed of the aeroplane, which slightly deflects the bullet in relation to the gun, when the gun is not firing straight ahead. This corresponds to the correction given by the windvane foresight when open sights are used. The telescopic sight has to be tilted mechanically and requires a special mounting and gear for this purpose. These mountings and gears are not yet issued for service use; until they are, these telescopic sights should be used for fixed guns only.

9. The ring seen in the sight forms a guide as to how much to aim off for the speed of the target aeroplane, the target aeroplane being placed in the ring in such a way that it will fly into the centre of the ring by the time the bullet reaches it. This is done exactly as though using a bead and ring sight. The telescope in fact, when fixed, exactly replaces the bead and ring sight used with fixed guns, and when carried on a compensating mount, replaces the ring and windvane sight for movable guns. The only difference is that the gunner is released from the

necessity of holding his eye aligned on two sights or of keeping his head at a fixed distance from the ring. So long as he can see the ring at all it is in the correct position for aiming.

10. The training, therefore, with the ring sights is exactly applicable to the Aldis, except that those parts of the training relating to the aligning of the eye and fixing of the head, are no longer necessary.

Fixing of the Sight.

11. For fixed guns, firing straight ahead, the telescope is held in a pair of brackets which are adjustable for line. So far as possible the fixing of the telescope is so arranged that, once adjusted, it remains correct. But it is necessary to check it from time to time. This can be done in the usual way by looking through the barrel of the gun at a distant mark with the aid of a mirror, and seeing that the mark falls in the centre of the ring.

12. The sight should be removed when not in use, by loosening the wing nuts on the clips attaching it to the brackets.

Part V.—INSTRUCTIONS FOR USING MODEL AIMING AEROPLANE.

(*See* Fig. 3.)

Description of Model.

1. This consists of a model of an aeroplane to 1-20th or 1-10th scale, and is intended for practice in aiming off to allow for the enemy's speed, and for judging range to determine the correct moment for opening fire.

2. The model is mounted upon a universal joint and carries a rod, sliding in its body.

3. On the end of this rod is a ball, which forms, when properly adjusted, the correct point of aim when allowing for the aeroplane's speed, no matter from what point of view the aeroplane is observed.

To set up the Model.

4. Set up the model and arrange practice guns and sights around it at the required range, so that the model appears against the sky or some uniform light background.

5. The actual range should be 1-20 or 1-10 the real range, *i.e.*, if working at 200 yards, the gun, when using a 1-20 model, should be 10 yards from the model ; other ranges in proportion.

6. Set up the guns on some form of mount such that they can be moved easily on their pivots but remain wherever put (*i.e.*,

the pivots should be very slightly stiff and the guns balanced upon them).

7. It may be found convenient to use dummy wooden guns for this purpose.

8. If the training is for a ring and windvane sight, the windvane must be replaced by a fixed bead before beginning practice; if for an Aldis optical sight, the sight must be fixed rigidly to the gun.

9. If the training is to be on the assumption that the speed allowed for corresponds to the ring on the sight provided, it is now necessary to draw out the stick from the body until, with the model flying directly across the line of sight and the centre of the ring on the bead, the vital part of the model is on the edge of the ring. This can be done either by trial on the spot—*i.e.*, by drawing out the rod until, when using the sight correctly, the above is true—or by having the rod marked to correspond with various ranges at the given speed.

10. If an advanced training including estimation of increased speed due to diving, etc., is required, the rod must be marked and adjusted accordingly, using the fact that at any range the distance between the ball and the vital spot in inches should be $0.024 \, V \times R$. Where V is the velocity in miles per hour and R the actual range in yards.

11. In any case it is of course necessary to readjust the position of the sliding rod whenever the range is altered.

To Use the Model.

12. The first training consists in turning the aeroplane to various positions and making pupils aim at the ball and note the appearance and position of the aeroplane in the ring. After some practice at this, remove the rod altogether and allow pupils to guess the correct position of the aeroplane in the ring, corresponding to its appearance; then check their aim by replacing the rod and observing how nearly the ball falls upon the centre of the ring.

13. The rod should always be removed entirely and not merely pushed right into the model, as if this is done the tail of the rod, protruding behind the tail of the body, unfairly assists the gunner in aiming so that the body points towards the centre of the ring.

14. The photo herewith shows the appearance of a 1-20 scale model at 10 yards corresponding to 200 yards on the full scale The distance between the ball and the centre of the ring gives the error in setting the sight.

15. This practice should be carried out at several different ranges, less than the equivalent of 300 yards on the full scale, and should be continued until the pupils can get the ball within the

Fig. 3.
DIAGRAM SHOWING USE OF AIMING MODEL.

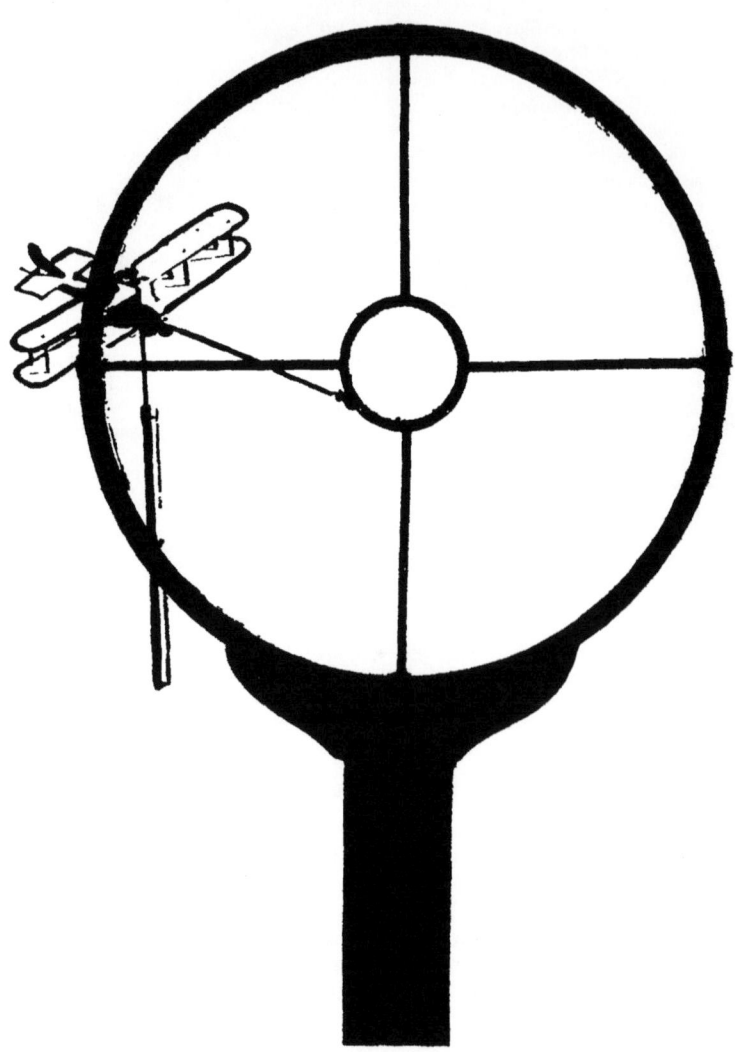

inner ring practically every time, whatever the position of the aeroplane and whatever the range.

16. In this way the laying of the aeroplane on the ring in the correct position, having regard to its appearance and not to range, should become instinctive; in addition the pupil should have become familiarised with its apparent size in the ring at the correct range for opening fire.

17. The training should not, at least in the first place, be complicated by any allowance for speed differing from those for which the rings are intended; attention should be concentrated solely on the two points; opening fire at the correct range and placing the aeroplane correctly in the ring.

18. In the early trials plenty of time should be allowed to lay the gun, but later this should be cut down to two and even one second.

19. It is very important to vary the range from time to time to prevent the pupil forming the habit of estimating his allowance in body lengths, a thing which he is very apt to do, and which is incorrect.

Part VI.—INSTRUCTIONS FOR USING THE HYTHE GUN CAMERA : Mk. II.

General.

1. The gun camera has been designed with a view to improving aerial gunnery. With the aid of this instrument it is possible to check the aiming or laying of the gun and to obtain graphic records of results: in other words photographs are produced showing where the bullets would have gone had the gun actually been used.

Instructions for Setting Up.

2. Care must be taken before practice to ascertain that the setting of the screen in the camera coincides with the sight on the gun.

3. The method of doing so is as follows :—The gun is mounted on a tripod with the camera (focal plane type Mk. II) clamped on the right hand side of the barrel casing. Take out the back of the camera and the metal film holder, and open the lens by turning the film changing handle slowly clockwise until it nearly reaches the lock, when the capping shutter will be heard to fly back. At this point stop the handle and release the shutter; then turn the handle slowly backwards until a certain resistance is felt, when it will be found that the screen is exposed enough for sighting.

4. Align the gunsights correctly on a sharply defined target some 100 to 200 yards distant; then set the camera so that the centre of the image of the target coincides with the centre mark on the screen. Lateral movement is obtained by moving the screen itself; vertical movement by adjusting the screw on the top of the screen-holder. When the target is correctly centred on both gun and camera, continue turning the handle in a clockwise direction, moving the lock in order to allow the handle to pass.

Use of the Gun Camera.

5. The gun camera is used in practice aerial duels, and it determines whether the gunner has succeeded in laying his gun correctly upon the enemy aeroplane. In actual shooting it is necessary, when laying the gun, to make allowance for the speed of the enemy aeroplane, and the gun camera shows whether this allowance has been correctly made. It cannot, however, be used to check the allowance, necessary in real shooting, for the speed of the gunner's aeroplane; therefore, as this speed is allowed for in all modern sights, it is necessary to remove whatever automatic allowance is provided on the sights in use on the gunner's aeroplane. With a ring and windvane sight this is done by replacing the wind vane sight by a fixed bead, and with the Aldis compensating sight the alteration is made by removing the compensating gear and fixing the telescope parallel to the axis of the camera.

6. To use the gun camera the gunner lays the sights on the target aeroplane, and presses the trigger of the gun at the moment when he considers that he would hit his opponent if he were actually firing. The effect of pressing the trigger is to expose the film in the gun-camera, and the resulting photograph when developed and measured shows how far he was right or wrong in his aim.

7. The handle on the right hand side of the camera must be turned once between each exposure. About 11 exposures can be taken on each roll of film.

Instructions for Working Out Results.

8. The film when developed shows a photograph of the target aeroplane and, either a series of ruled lines dividing the film into squares, or a series of concentric circles, this depending on which type of screen is used. In either case the true centre is clearly shown. (The screen should have been previously aligned with the gun sights.) (*See* para. 4.)

9. It is possible from this photograph to determine where a bullet, fired when the trigger was pressed, would have hit or passed the aeroplane had it been flying at any given speed and had an accurate automatic sight to allow for the gunner's own

speed been used. In other words it is possible to tell whether the gunner has carried out his part of the shooting correctly.

10. To do this proceed as follows :—

11. Scratch a line on the film through the body of the aeroplane parallel to the direction of motion (in a B.E. this line passes through the propeller boss and the front edge of the fixed tail plane where it meets the body).

12. From the centre mark on the film scratch another line parallel to the above line, and measure along this second line from the centre mark, in a direction opposed to that of the flight of the aeroplane, a distance corresponding to the allowance for the target speed. This distance can be found approximately by methods to be described later.

13. The point so found, called hereinafter the point A, is the spot where a single bullet, fired when the trigger was pressed, would have passed the target ; the distance between this point and the spot aimed at on the picture of the aeroplane represents the gunner's error in laying the gun.

14. If a series of real shots had been fired they would have been spread over a fairly wide area around this point.

15. A circle of 0.22 inches radius drawn about A as centre will represent the spread of a drum of bullets fired by an average gunner under average conditions. If the vital parts of the aeroplane do not lie within this circle it is very improbable that they would have been hit ; if, however, they do lie within it the chance of a hit depends on the number of shots fired and the range. Half a drum grouped within this circle at 200 yards range would represent an even chance of a hit. This applies when the target aeroplane is flying more or less across the line of flight. Where the target is flying straight towards the gunner, the chance is reduced because the vulnerable parts are in line and cover one another. On the other hand the shooting under these conditions is usually easier.

Method of Finding the Correct Allowance.

16. The distance of the point ' A " from the centre of the film depends upon the assumed speed of the target aeroplane and the angle at which it is crossing the line of sight.

17. The target may for this purpose be assumed to have any pre-arranged speed, which need not necessarily be that at which it was actually moving. This assumed speed will generally be that for which the ring sight, used by the gunner, was designed (*i.e.*, 100 m.p.h. with the sights now issued).

18. If the assumed speed is 100 m.p.h., and if the target is flying directly across the sight line, the distance of " A " from the centre of the film is 0.73 inches, or the radius of the largest circle, when using a camera which marks circles on the films.

19. If the target is flying towards or away from the camera

this distance must be reduced in proportion to the foreshortening of the body, and should a rough estimate only be required, can be guessed from the appearance of the aeroplane (*e.g.*, if the body appears to be foreshortened to half its natural length the distance of " A " from the centre is reduced by one half or to 0.37 inches at 100 m.p.h.

20. If a more accurate estimate of the allowance is required it can be obtained by reference to the 56 pictures (reproduced in Fig. 4) of a B.E. aeroplane taken from different aspects. These pictures are arranged in such a way that all in one column require the same allowance. so that it is only necessary to determine in which column the picture that most resembles the photo on the film lies, and then to note the allowance corresponding to this column.

21. To pick out this picture in the series it is necessary to consider from what point of view the actual photo was taken, and hence into which of the two groups of pictures it falls (*e.g.*, if it was taken from behind and below or in front and above it will fall in one group, and if from behind and above or in front and below, in the other). The comparison is then easiest made by rotating the film until the body of the aeroplane is parallel to the bodies of the key aeroplanes, and then noting the angle of the wings and position of the tail etc. In some cases it is necessary to reverse the film and look at it from the back·to simplify the comparison.

22. When once the column containing the picture most nearly resembling the photo has been decided, the distance of the point " A " from the centre of the film can either be read off in inches at the top or the bottom of the column, or measured by dividers from the lengths given at the bottom of the columns, or if the films are marked with concentric circles the required distance can be marked off without any measurement at all, since the numbers 20, 40, 60, 80, 100 beneath the column represent the radius of the five circles on the film ; thus if the chosen picture is in the column above the number 60 the allowance is equal to the radius of the 60 or third circle and so on.

23. Another way of looking at this matter is to consider the numbers beneath each column as representing the speed at which any aeroplane in the column is crossing the line of sight, assuming that it is travelling through the air at 100 m.p.h. The radii of the five circles on·the film represent the allowance which it is necessary to make, when using Mark VII ammunition, in order to hit any object crossing the line of sight at 20, 40, 60, 80 or 100 m.p.h. respectively.

24. The above operations, though lengthy to describe, can be performed very quickly ; with practice a few seconds only are required to spot the key picture which most nearly resembles the photograph, and hence to determine the proper allowance.

PHOTOGRAPHS OF "B.E." AEROPLANE FOR USE WHEN MARKING HYTHE GUN CAMERA.

FIG. 4.

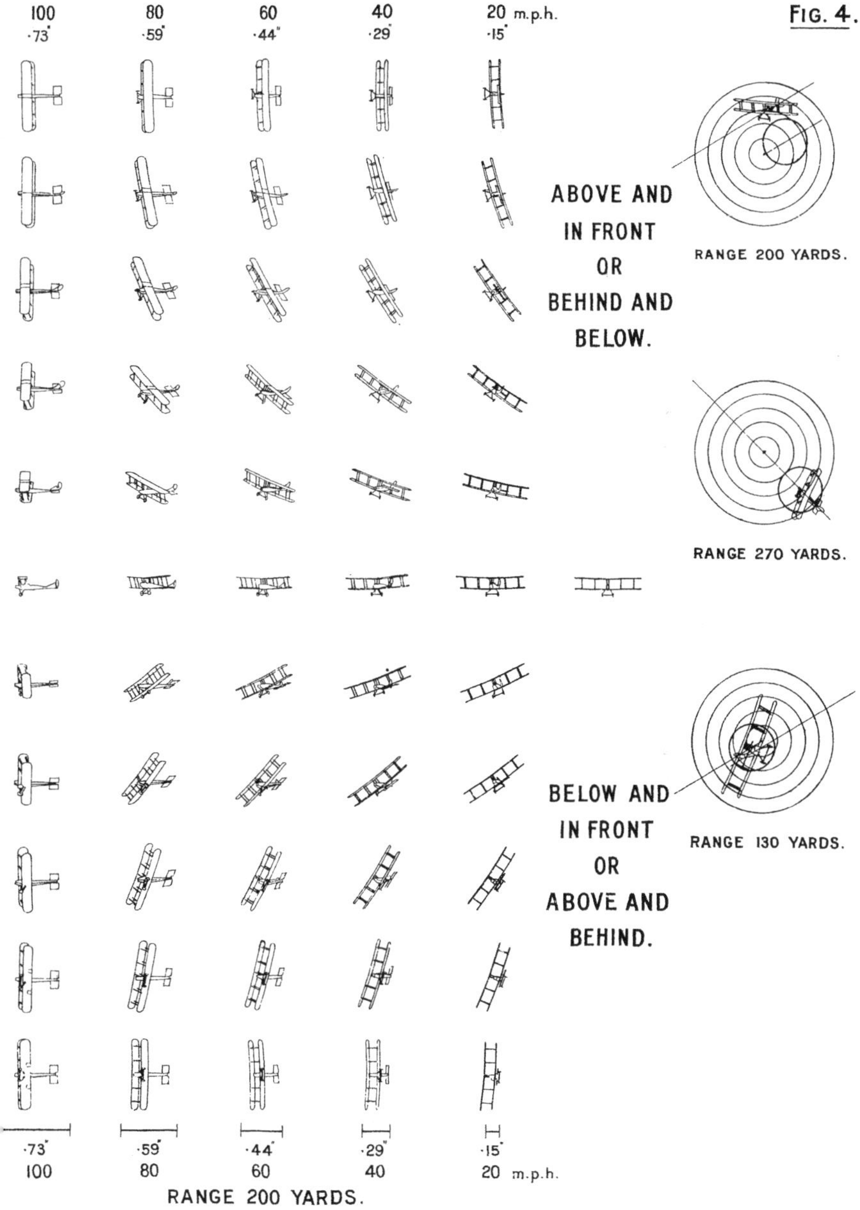
100 ·73"
80 ·59"
60 ·44"
40 ·29"
20 m.p.h. ·15"
ABOVE AND IN FRONT OR BEHIND AND BELOW.
RANGE 200 YARDS.
RANGE 270 YARDS.
BELOW AND IN FRONT OR ABOVE AND BEHIND.
RANGE 130 YARDS.
·73" 100
·59" 80
·44" 60
·29" 40
·15" 20 m.p.h.
RANGE 200 YARDS.

14654/9442. 3500. 3.17.

25. Should other speeds than 100 m.p.h. be assumed when using the camera, the allowances worked out must all be increased or decreased in proportion to the speed.

26. Distance of "A" from the centre of the film is independent of the range. If it is desired to find the range proceed as follows :—

27. Measure any well defined length of the photograph of the target aeroplane, *e.g.*, the apparent span, as foreshortened, and then measure the corresponding length in the chosen key photograph. The range in yards is then obtained by dividing the first of these lengths (photograph of target) into the second, and multiplying the result by 200.

28. If the target aeroplane is not a B.E., fairly accurate results can still be obtained from the key pictures if it is of the same type as a B.E., *i.e.*, tractor biplane, but if the span and body length are much different from those of a B.E. (*e.g.*, in a Bristol Scout) the range must be altered in proportion.

Note.

29. The actual key pictures were all taken from in front, no pictures taken from behind are shown because these would so far as outline and allowance is concerned, exactly produce the series taken from in front, except for the effect of perspective, which at these ranges is negligible.

30. An example of a graded series of practices with the gun camera is appended. These practices are in use at Hythe in the instruction course for aerial gunners.

(a) On the ground, gun on ordinary Lewis tripod.

Practice 1.—Pupil to be seated on the tripod with the gun camera in usual position on the gun. A stationary aeroplane is 200 yards away—front view.

The holder is taken out of the camera and the shutter opened. Each pupil is then told to aim directly at the centre of the machine, *i.e.*, (propeller boss) getting the circle and bead of ring sight in correct alignment with machine as though firing at it. The instructor checks his aim by looking through the screen of camera. After pupils have had a certain amount of practice a film is placed in camera and each pupil takes one aim at machine, presses the trigger and turns the handle one revolution.

In this practice the pupil's power of correctly aiming, correct use of sight, and also his ability to manipulate the camera, is observed.

Practice 2.—As for 1, except that the machine is placed at right angles to the firer. The holder is again removed and each pupil practises aiming at the machine as if it were flying at the speed represented by the ring on the gun, *i.e.*, he places the pilot of target machine on the edge of the ring, his aim being checked as in 1; the pupil then takes one photograph as in 1.

Practice 3.—As for 2, from the ground, but in this case the machine is flying across the front at about 500 feet. Pupil takes three consecutive aims at the machine allowing for deflection by means of the ring sight, placing the pilot of target machine on the edge of ring as in practice 2. In order to carry this practice out correctly it is essential that the target aeroplane maintains a straight course, flying at right angles to firer.

(b) Aerial practices.

Practice 4.—As for 1, but carried out in an aeroplane in the air. One machine is sent up as target, another machine (B.E. 2c.) is sent up containing a pupil armed with gun camera mounted on the Strange mounting. The target machine will manœuvre to follow directly behind machine with camera. The pupil then takes six consecutive aims, turning handle one revolution between each aim. In this practice the pupil must not open fire until he is within the limit range, *i.e.*, 100 to 300 yards. Marks will be allotted for correct aiming and ranging.

Practice 5.—As for 4, but the machines approach each other at various angles, and the gun is mounted on the side mounting. Each pupil takes six consecutive aims, placing the target machine within the ring sight according to its angle of approach.

Marks will be allotted as in Practice 4.

Practice 6.—Two pupils are sent up in separate machines and are armed with gun cameras. They are told to engage in combat, allowing for the movement, etc., of each other's machine exactly in the same manner as though actually fighting. Each pupil takes six aims as before and the results are checked by :—

(*a*) The range at which fire was opened.
(*b*) Position of machine in ring.
(*c*) The angle at which the opposing machine was engaged so as to obtain a blind spot.

Part VII.—SHOOTING PRACTICE AT A PICTURE TARGET.

Object of Practice.

1. When shooting in the air with any ring sight, whether it be of the ring and wind vane type, or of the Aldis optical type, it is generally necessary to aim off the target aeroplane so that it appears in the ring in some particular position depending upon its aspect. The ability to estimate the correct position corre-

sponding to any aspect can best be acquired by constant practice at the aiming model, as described in Part V. Even when this knowledge has been attained, considerably more practice is required before a man can hold a firing gun on to a moving target and maintain continuously the correct allowance.

2. For this reason it is important that some form of practice on the ground shall be available which allows a gunner, already trained upon the aiming model, to put this experience into practice with an actual Lewis gun firing at a picture of an aeroplane. This practice must be so arranged that the concentration of his group and the correctness of his allowance can be easily checked, and so that the gunner does not remain fixed on his stand while shooting, but has to swing his gun in the same way as he would have to do in the air in order to follow the moving aeroplane.

3. Since the movement of gunner and target in the air is only relative, it is immaterial whether in the ground practice the target or the gunner is moved. For instance, if two machines are flying parallel and at the same speed, each will appear stationary to the other. This is represented on the ground by having a fixed picture of an aeroplane on a reduced scale broadside on, the gunner firing from a fixed point at a range reduced in proportion to the scale of the target so that the aeroplane has the same apparent size as on full scale. He is provided either with a fixed bead foresight and with ring backsight, or with an Aldis telescope fixed parallel to the axis of the gun. In either case he aims ahead of the picture in the direction of its apparent flight, making the correct allowance, and his bullets make a group somewhere ahead of the picture.

4. Again, the case of the target aeroplane flying directly at the gunner, but at right angles to his line of flight, so that he is shooting over the port side of his machine, the target will appear to move from right to left, that is, it will appear first well ahead of the gunner's machine on the port bow and as he flies on will gradually move astern. To follow it the gun must be swung from right to left while firing.

5. Any relative movement of gunner and target may be represented on the ground by fixing the target and putting the gunner on a turntable and slowly turning him round. The effect of this is to make the target pass across his field of fire, and in shooting at it he has to swing his gun so as to keep it bearing on the target, exactly as he would do in the air. Practice under these conditions will enable him while swinging the gun to keep his sights on the target, always with the correct allowance.

6. In order to correspond with conditions in the air it is important that both the range and aspect of the target should be constantly varied. This may be done by having a large number of pictures of aeroplanes in different aspects, ranging from flying

straight towards the gunner to flying straight across him, and by making him fire at these targets at different ranges.

7. A range with which this practice can be carried out can be very simply made as follows.

Construction of a Suitable Target and Range (*see* Figs. 5, 6 and 7).

8. The target, which is fixed, consists of a wooden frame about 15 by 12 feet, covered with light canvas (Fig. 5). In front of this screen, and touching it, is hung a light batten frame A, covered front and back with stout millboard or three-ply wood or other similar material, the frame being supported by cords D D attached to two corners and slung over the top edge of the screen; these cords carrying balance weights at the other ends.

9. Other cords may be attached to the lower corners of the frame and fastened to hooks at the foot of the screen; these prevent the frame from blowing about with the wind.

10. Several sheets of paper showing pictures of aeroplanes in various aspects to 1/10th scale are required. These pictures show the aeroplane from different points of view, and the more realistic they can be made the better, because the gunner should have every legitimate aid possible to enable him to estimate rapidly the aspect at which the picture aeroplane is presented to him.

11. These pictures are pasted on the millboard of the target frame, and should be so placed that, if the gunner shoots with the correct allowance, the bullets will strike on the canvas part of the screen. Care should be taken not to place the picture so that the body of the aeroplane is parallel to the edges of the frame, as this unfairly assists the gunner in estimating the direction in which the body is pointing. If a number of target frames are prepared each carrying two pictures, one on either side, they can be very quickly changed, and thus the aspect and direction of flight of the target can easily be varied for each shoot.

12. The bullets fired will then show upon the canvas screen, and the accuracy with which the allowance has been made can be rapidly checked by a method explained later. If the screen be whitewashed from time, to time, all old bullet marks will be obliterated, and new ones can be easily distinguished.

13. The gunner fires from a gun mounting, preferably a Scarff mounting, fixed in a mock-up of part of a nacelle which is carried on a turntable (A in Fig. 6). An ordinary contractor's 2 ft. gauge railway turntable does very well, but the surface on which it runs should be good so that the gunner can be turned round smoothly and without shake. It is best to mount the turntable 3 or 4 feet from the ground, so that the gunner fires slightly downwards, as by this means the size of the butt behind the target may

To face page 24.

Fig. 5.

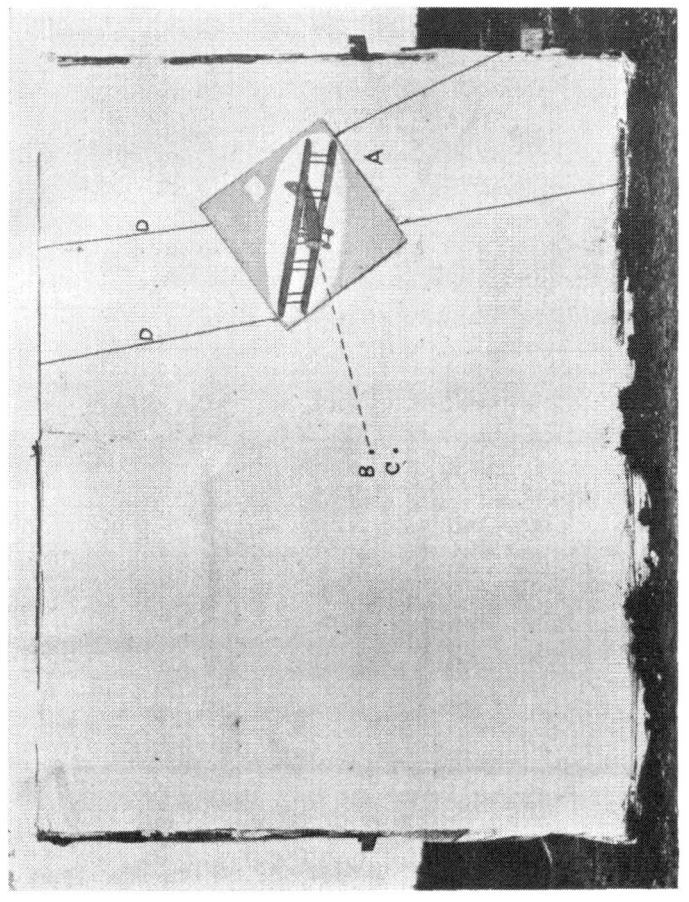

Fig. 6.

Following Fig. 5.

be reduced. Attached to the turntable is a lever by means of which a man can turn the gunner round, as shown in Fig. 6.

SKETCH OF ARRANGEMENT OF TURNTABLE NACELLE FOR TARGET SHOOTING PRACTICE.

Fig. 7.

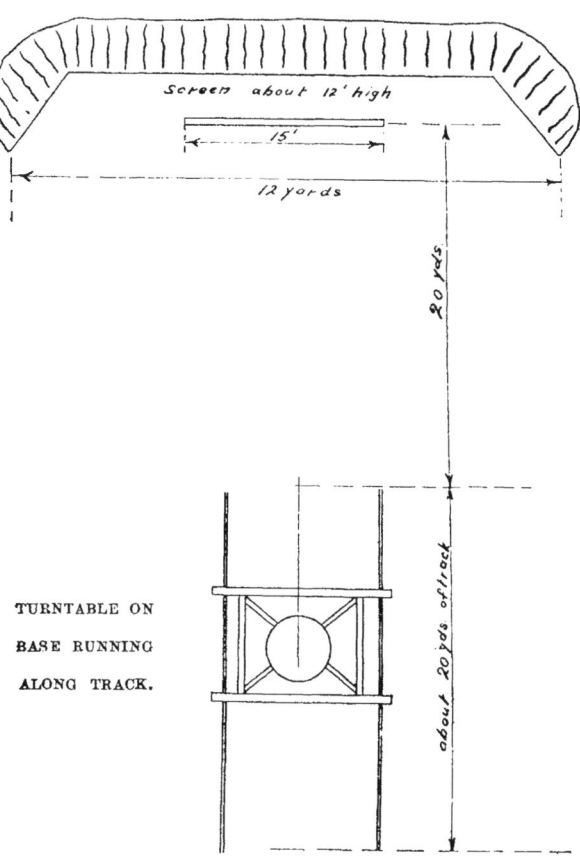

14. It is a good plan to mount the turntable and its stand on wheels, running on rails or otherwise, so that it can be easily moved a few yards towards the target. These should have a sufficiently wide base for the platform to be quite steady when the nacelle is being turned. By this means the distance of the gunner from the target, which it is important to vary constantly

in order to correspond to the varying ranges which occur in actual practice, may conveniently be changed between the shoots from say 10 to 30 yards. By turning the gunner and at the same time moving him along the track during the shoot, the conditions are made still more realistic ; if this is done, however, the marking becomes more difficult, as the correct grouping centre varies during the shoot. This system of mounting the turntable on a truck is not shown in Fig. 6, but it is shown in Fig. 7, which is a dimensioned plan of a range constructed according to this system. The butt behind the target will be of such a height as is necessary for safety.

15. In using this range the gunner is first turned away from the target so that he cannot see which silhouette is being put on the target. One of the silhouettes is then hung up in such a position that the bullets, if correctly aimed, will fall on the canvas screen. The turntable is then slowly turned round, until the target comes into the view of the gunner, who begins to shoot as soon as he can bring his gun to bear, putting on the correct allowance by means of the ring sight in accordance with the aspect of machine shown on the silhouette. His rotation may be continued right round during his shoot, or may be stopped and reversed, and then possibly again reversed and moved ahead. The rotation of the turntable at any moment may be in any direction and at any speed up to the rate of two revolutions per minute, corresponding to the maximum rate at which he is ever likely to have to traverse his gun in practice.

16. In this way the gunner can be suddenly presented with a target of unknown aspect at an unexpected range, and he will thus have to decide the necessary allowance and fire his shots correctly, in the time taken by the target to cross the field of fire.

17. The silhouettes provided for this practice are all of 1/10th scale. Consequently a distance of 20 yards on this range corresponds to 200 yards in actual practice. Also, the allowance on the target is 1/10th of the allowance for the movement of the target aeroplane at the corresponding range on a full scale. If the gunner has fired correctly at the picture his bullets will not have struck the picture itself, but will have hit the canvas screen at a point ahead of it equal to this distance.

Method of Checking the Accuracy of the Allowance.

18. To find the spot which should have been hit, proceed as follows :—

19. Lay a straight edge, or a stretched string, along the axis of the body of the picture, and on this line mark off a point " B " at a distance in front of the vital part of the picture corresponding to the correct allowance (*see* Fig. 5). This is the point on which the centre of the ring should have been laid if the gun were aimed correctly.

20. The actual bullets would, if correctly fired, strike about 6 inches below the point " B " because the barrel of the gun is about 6 inches below the sight line. Mark off, therefore, a point " C " 6 inches below " B." This is the point where the bullets would have struck had the gun been correctly aimed.

21. The photograph of the target, Fig. 5, shows a silhouette in place, and the points " B " and " C " marked on the canvas. These points should not, of course, be visible to the gunner in actual practice.

22. The distance of the point " B " in front of the vital spot depends upon the ring sight used, the range, and the apparent angle of approach of the picture target. With a 100 m.p.h. ring sight, and a picture of an aeroplane flying directly across the screen, the distance is almost exactly one-fifteenth of the range, *i.e.*, at 20 yards this distance is equal to 4 feet and at 30 yards it is 6 feet. For a foreshortened photograph the distance is reduced in proportion to the foreshortening of the body, *i.e.*, if the aeroplane is shown flying at 30 degrees to the line of sight, the distance is halved.

23. A series of poster pictures of aeroplanes seen from different aspects will be issued for this practice. These pictures will carry two marks to give the correct line of flight, and a mark in the centre of the vital region. Each picture will also carry a table showing the appropriate allowance at different ranges. Until these can be provided, however, silhouettes can easily be made locally.

24. A perfect shoot at the target should show a small group with its centre at C. If it is desired to ascertain whether a vital hit would have been scored, prick off the shots and also the point C on a piece of thin paper and then transfer the paper, keeping it parallel to itself, so that the point C on it lies over the pilot's seat. The pricked shots will then show where the aeroplane would have been hit in real shooting.

Or :—Prepare beforehand outline drawings on tracing paper, one for each of the poster pictures. After the shoot has been made, and the point C found, place the tracing drawing corresponding to the picture target, with its centre of the vital region on the point C and its body pointing in the same direction as the body of the aeroplane on the picture. This affords a quick check upon the gunner's shooting.

Part VIII.—NOTES ON THE USE OF VARIOUS PRACTICES.

Necessity for Ground Practices.

1. In most air fighting the gunner, before actually firing, has to get his gun to bear upon the enemy aeroplane, decide whether the opportunity is worth the expenditure of ammunition, and estimate and make the necessary allowance for the enemy's speed. In a close fight all these things have to be done with the greatest rapidity if any effective shooting is to be got in at all, and the gunner is generally working under very considerable difficulties owing to the diving, banking, &c., of his own aeroplane.

2. For this reason no operation which cannot be performed almost without conscious thought will be of much value in a close fight. The gunner requires all his thinking powers to attend to the general features of the fight and has practically none to spare for the details of aiming.

3. It is generally found very difficult to get anything like sufficient actual flying to develop the required instincts to the high pitch necessary for accurate air shooting, under actual fighting conditions, and it is therefore essential that every air practice should have its counter-part upon the ground, where an almost unlimited time may be spent in practice. These ground practices should, as far as possible, teach exactly the same operations as are required in the air, so that the gunner, when he gets his necessarily limited air practice, may not have to learn anything new, but may give his whole attention to applying the skill he has obtained on the ground to the more difficult and uncomfortable conditions necessarily met in the air, and may have his mind free to consider such questions as the extra allowance necessary for possible variation in the enemy's speed due to diving, climbing, &c.

Distinction between Aiming and Grouping.

4. For a man to become a good air gunner he must acquire skill of two kinds :—

(1) Skill in estimating the necessary allowance for the enemy's speed.
(2) Skill in holding the gun steadily in relation to a moving target despite the recoil of the gun and the oscillations of the aeroplane that carries it.

These can be taught separately at first and eventually combined when the pupil has become familiar with the methods of allowing for enemy speed and is able to hold his gun fairly steadily on a target when rapidly traversing his gun.

5. The following is a possible scheme of practices to achieve this end. The apparatus necessary for all these exists and is in constant use either at Hythe, Orfordness or elsewhere.

6. The reason that so many different practices are included in the scheme is that it is impossible to devise any single practice that will combine all the requirements necessary for good shooting under actual fighting conditions. This can only be obtained in a real fight where both combatants stand a great chance of being killed. The essential parts of the operations necessary are all contained in the following practices, but each individual practice necessarily leaves out some essential to real fighting. The notes on each practice, which follow the table, are intended to indicate roughly the reason for the practice and the deficiencies in it which make the other practices necessary.

7. Table of Practices for Training with Ring Sights.

For teaching aiming off without shooting.

| On ground | Aiming Model. See Part 5. |
| In air | Gun camera. See Part 6. |

For teaching shooting without aiming off.

| On ground | Practice aiming direct at fixed targets, from a revolving mounting. |
| In air | Shooting from aeroplane at large fixed targets using compensated sights. |

For teaching shooting and aiming off combined.

| On ground | Picture target. See Part 7. |
| In air | Shooting at flag targets from an aeroplane. |

Aiming Model.

8. This shows quickly and simply the reason for the allowance in shooting, and allows a very great amount of practice at judging the allowance to be obtained in a short time. Several pupils can work at once with one model and instructor, and all their shots can be rapidly checked.

9. The deficiency in this practice is that it does not give the idea of judging a constantly changing allowance on a moving object. It should therefore be considered as giving only the ground work for training in aiming off. It does not at all follow that a man who can aim off correctly at the model will be able to do the same in the air, without considerably further practice.

Gun Camera.

10. This provides a means of checking aiming in the air under fighting conditions. Practice under these conditions is essential as the difficulties, both physical and mental, of bringing the gun to bear on the enemy with the correct allowance are very great, even to a man who knows at a glance what allowance should be made under all conditions. It is also essential that a man's performance under these conditions should be checked, as it is difficult even after considerable experience to tell whether one has judged one's allowances and ranges correctly, and the results on the films are often very surprising, especially those taken in a vigorous fight.

11. The elements lacking in this practice are the necessity of holding the allowance correctly on a moving target, and the recoil of the gun. It is very much easier to pick an opportune moment when the allowance is correct and expose one film, than to hold the gun steadily and correctly on a moving target. It is hoped to produce a cinematograph camera for this purpose, but this is not yet available.

Direct Ground Shooting.

12. This is no doubt necessary for early training and is comparatively easy to provide. It should be remembered, however, that shooting from a fixed mount at a fixed target is very unlike shooting in the air, as the gunner can in this case steady himself and set his muscles in a way which he cannot do in air fighting, where the enemy is always moving relatively to himself. Quite a slow rate of revolution of the mounting is sufficient to make the gunner keep his muscles on the move and makes a very great difference to the size of group obtainable.

Air Shooting at a Fixed Target.

13. This should always be done at a large target using compensated sights. The object is to teach the gunner to get a good concentrated group aiming directly at an object from the air, and to give him confidence that his compensating sight really does allow approximately for his air speed. The target must therefore be large or the concentration of the group cannot be observed, the number of hits on a small target in any one shoot being largely a matter of luck, even if the shooting is good.

14. Quite useful work can be done with a large target lying on the ground, but if a large flag target suspended from a kite balloon can be obtained, the shooting can be made much more realistic, the aeroplane passing below and on a level with the target, as well above it.

15. The defect of this practice is that the relative movements of the target and the aeroplane are very different from those of

two aeroplanes in a fight. The gunner therefore does not get any useful experience of shooting in a fight, but merely checks the accuracy of his aim and of his own compensated sight when in flight.

Picture Target Practice.

16. The object of this practice is to combine aiming with allowances, and firing while traversing the gun. Even although a gunner may know exactly what allowance to make for any given appearance of an aeroplane, and may be able to make a good group with his gun, a certain amount of practice is necessary before he can combine the two. If a ring and bead sight is used the practice is also of value in training the gunner to hold the bead in the centre of the ring whilst at the same time observing that the aeroplane is correctly placed in the ring.

17. This practice has the merit that the gunner has to estimate his allowance from the appearance of the picture, and has to put his estimation to immediate effect by actually firing the gun, exactly as he would in the air; the accuracy with which he has done this can then be checked very easily and quickly.

18. The defect of the practice is that a certain amount of imagination is required to see the silhouettes or pictures as though they were approaching or receding from the observer, and it is for this reason that it is important to have good pictures where possible. The practice of course gives no experience of actual fighting conditions but it is intended merely for training the gunner in the actual operations of aiming and firing. Fighting experience to supplement this practice is obtainable with the gun camera. A gunner who can do well both at the picture target and with the camera in a sham fight should certainly prove to be a good shot when he comes to a real fight.

Air Shooting at a Towed Flag.

19. This practice is valuable in that it provides the only chance the gunner gets of firing at a rapidly moving target from an aeroplane, and thus gives him confidence that his previous training, hitherto rather unreal, has in fact enabled him to hit a rapidly moving target in the air from a rapidly moving aeroplane.

20. One defect of this target, however, is that it does not look like an aeroplane, and gives very little indication of its angle of approach, so that the gunner's previous training in estimating the angle of approach of an aeroplane is of little use to him. Another defect is that the area presented to the gunner becomes very small when the target is flying towards or away from him.

21. For the above reasons it is only practicable to fire at the flag when it is moving across the field of view, so that the gunner

gets little or no practice at judging allowances and very little experience of actual fighting conditions.

Notes on other Forms of Practice.

22. Other practices such as aiming at clay pigeons, shooting at disappearing aeroplanes, &c., may be valuable to train the gunner's general powers of quick decision, but bear no other definite relation to the problem of shooting in the air. Although these may be useful to keep up the gunner's interest in his work, they should not replace practices, such as those outlined above, which teach a necessary part of air fighting.

23. Shooting directly at a small travelling aeroplane, unless the aeroplane always points directly at the gunner, is apt to be misleading, because this is the one thing that never should be done by a gunner using compensating sights.